Mathematical Olympiads

for

Elementary School 3

My First Book of Mathematical Olympiads – *Third Grade*

(Workbook Plus)

My First Book of Mathematical Olympiads

Mathematical Olympiads *for* Elementary School

3

Third Grade

(Workbook Plus)

Michael Angel C. G., Editor

Preface

The Mathematical Olympiads for the Third Grade of Elementary School discussed here are none other than the *Mathematical Olympiads for Schoolchildren "Unikum"*, which are held every year in the city of Lipetsk since 2010, and organized by the Faculty of Physics, Mathematics and Computer Science of Lipetsk State Pedagogical University and the Center for Continuing Education of Children "Strategy". Likewise, these Olympiads consist of two rounds, a qualifying round and a final round, both consisting of a written exam. The problems included in this book correspond to the final round of these Olympiads.

The present edition called Workbook Plus seeks to consolidate the mathematical skills acquired with the previous workbook since it includes new variants of problems as well as more challenging tasks. In this workbook has been compiled all the Olympiads held during the years 2011-2020 and is especially aimed at schoolchildren between 8 and 9 years old, with the aim that the students interested either in preparing for a math competition or simply in practicing entertaining problems to improve their math skills, challenge themselves to solve these interesting problems; or it could even be used for a self-evaluation in this competition, trying the student to solve the greatest number of problems in each exam in a maximum time of 1 hour 10 minutes. It can also be useful for teachers, parents, and math study circles. The book has been carefully crafted so that the student can work on the same book without the need for additional sheets, what will allow the student to have an orderly record of the problems already solved.

Each exam includes a set of 8 problems from different school math topics. To be able to face these problems successfully, no greater knowledge is required than that covered in the school curriculum; however, many of these problems require an ingenious approach to be tackled successfully. Students are encouraged to keep trying to solve each problem as a personal challenge, as many times as necessary; and to parents who continue to support their children in their disciplined preparation. Once an answer is obtained, it can be checked against the answers given at the end of the book.

Sincerely,

The editor

Contents

Problems

Olympiad 2011

(II Mathematical Olympiad "Unikum")

Problem 1. To cut an iron pipe into 2 parts, 50 rubles must be paid for the service. How much will the service cost if the tube needs to be cut into 12 pieces?

Problem 2. It is necessary to arrange 90 bags in three boxes so that the first box contains twice as many bags as in the second, and in the second there are 2 bags more than in the third. How many bags will there be in the first box?

Problem 3. The weight of a box with oranges is 35 kg. After half of the oranges were sold, the box was put on the scale. The scale showed 21 kg. What is the weight of the empty box?

Problem 4. In some state there are five cities: Multiplication (M), Division (D), Addition (N), Subtraction (S) and Arithmetic (A). There is a road between every two cities. Sasha lives in the city of Arithmetic and wants to visit each city in one trip, and then return back to the city of Arithmetic. Give all possible routes for Sasha, but those in which Sasha has not visited any city (except for the city of Arithmetic) twice. How many different routes can be found?

For example, routes where the first, after the city of Arithmetic, is the city of Multiplication: AMDSNA, AMDNSA, AMSNDA, AMSDNA, AMNSDA, AMNDSA.

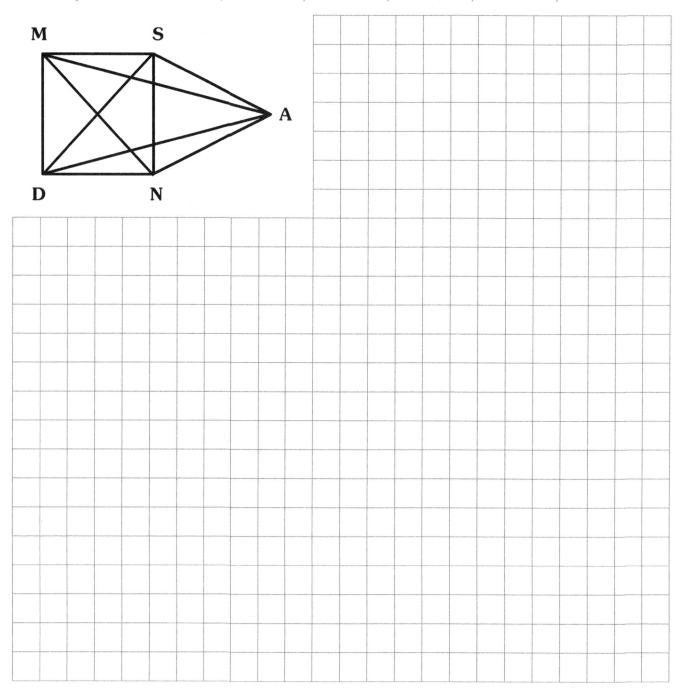

Problem 5. The fourth-grader decided to multiply all the days of the month of May, from the first to the 31st, and then subtract his lucky number 2011 from the result and determine the last four digits of the final result. The fourth-grader puffed and tried for a long time. So he would have thought until the evening, but his friend Max came and quickly solved the problem. Offer your own solution to this problem.

Problem 6. Can the shape shown in the figure be cut into four equal parts along the edges of the cells so that they can be assembled into a square? If possible, show one of the ways to get the square. If not, can you explain why?

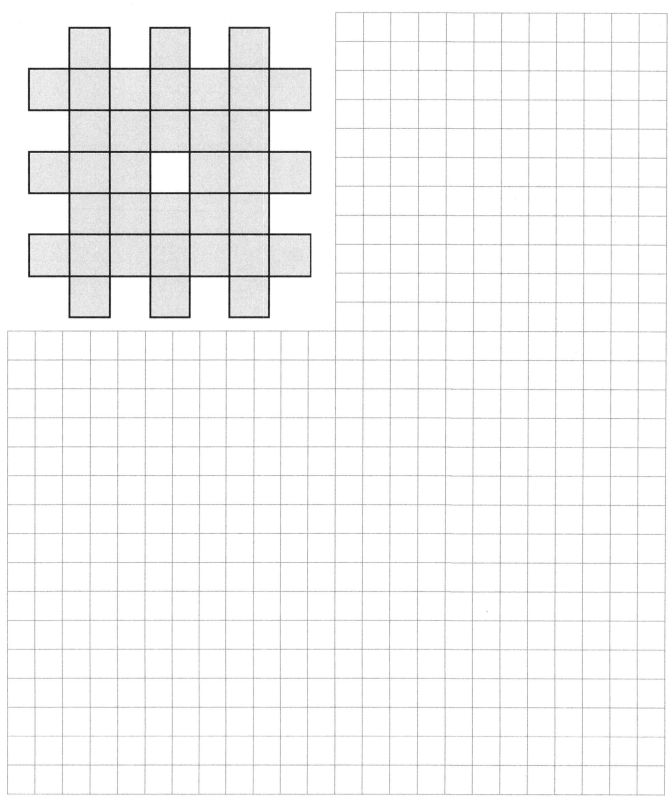

Problem 7. The numbers 1, 2, 3, 4, 5, 6 are written on the faces of a die. The die is rolled twice. In the first time, the sum of the numbers on the side faces is 13, and in the second time it is 16. What number is written on the opposite face to the one on which the number 4 is written? Explain why there are no other solutions besides the one found.

Problem 8. Find all the solutions to the number puzzle

UNIKUM + UNIKUM + UNIKUM + UNIKUM + UNIKUM +

+ UNIKUM + UNIKUM + UNIKUM = REGINA

If different letters represent different digits. Explain why there are no solutions other than those found.

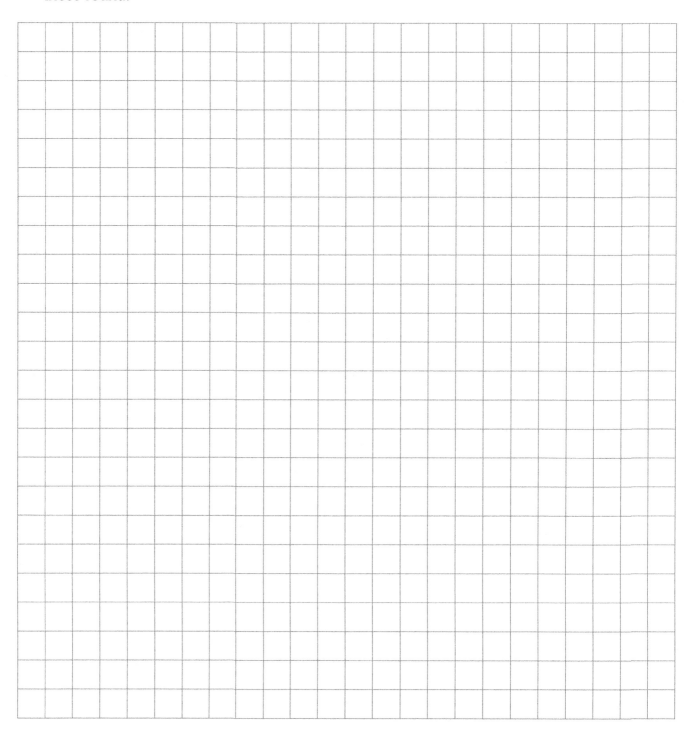

Olympiad 2012

(III Mathematical Olympiad "Unikum")

Problem 1. ¿ What digit does the product of 50 factors end at,

$$2111 \times 2113 \times 2115 \times \ldots \times 2207 \times 2209?$$

Problem 2. Three students have together 12 books with entertaining math problems. The first student has two less books, and the third student has two more books than the second student. How many books with entertaining math problems does each of the students have?

Problem 3. A bag contains 40 pencils. Eight of them are red, 7 are yellow, 25 are blue. What is the largest number of pencils you can take from this bag with the eyes closed so that at least 4 pencils of the same color and at least 3 pencils of a different color remain in the bag?

Problem 4. A bacterium was introduced into a flask. The number of bacteria doubles every minute. After three hours, the flask was filled with bacteria. At what point was a quarter of the flask filled with bacteria?

Problem 5. There are three boxes: a box of oranges, a box of apples, and a box of a mixture of apples and oranges. Each box has a sign indicating what is inside. The signs were taken and mixed; now it turned out that all the signs were out of place. You have one attempt: you can put your hand in a box and pull out one fruit. After that, you need to hang up the signs correctly.

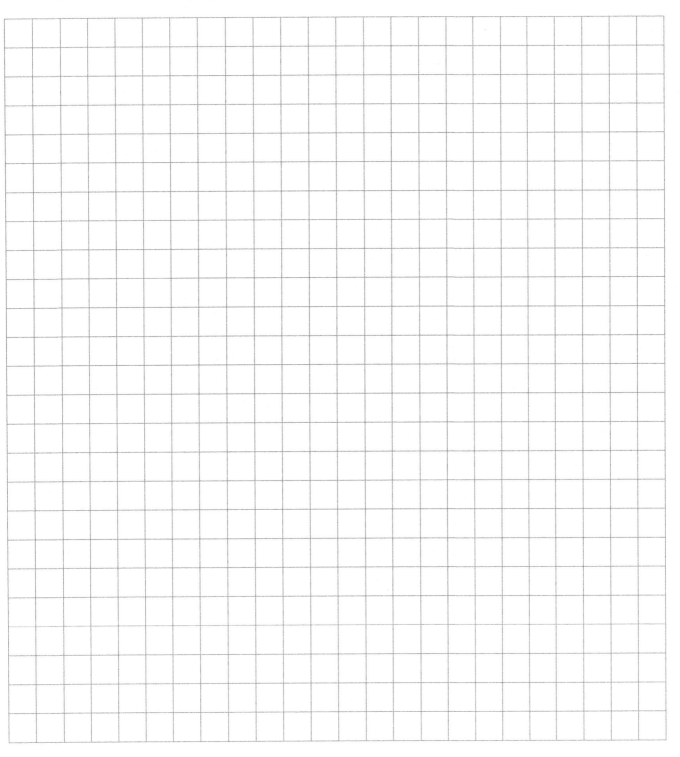

Problem 6. Is it possible to cut the shape shown in the figure into four equal parts (identical in shape and size). If so, how?

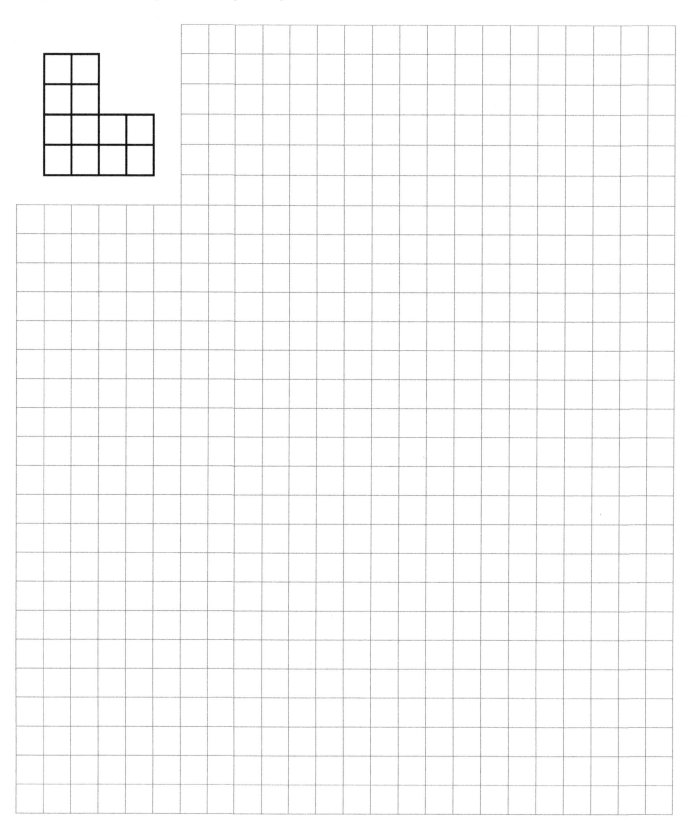

Problem 7. Find all the solutions to the number puzzle:

$$MA + TE + MA + TI + KA = UU.$$

If different letters represent different digits. Explain the lack of solutions other than those found.

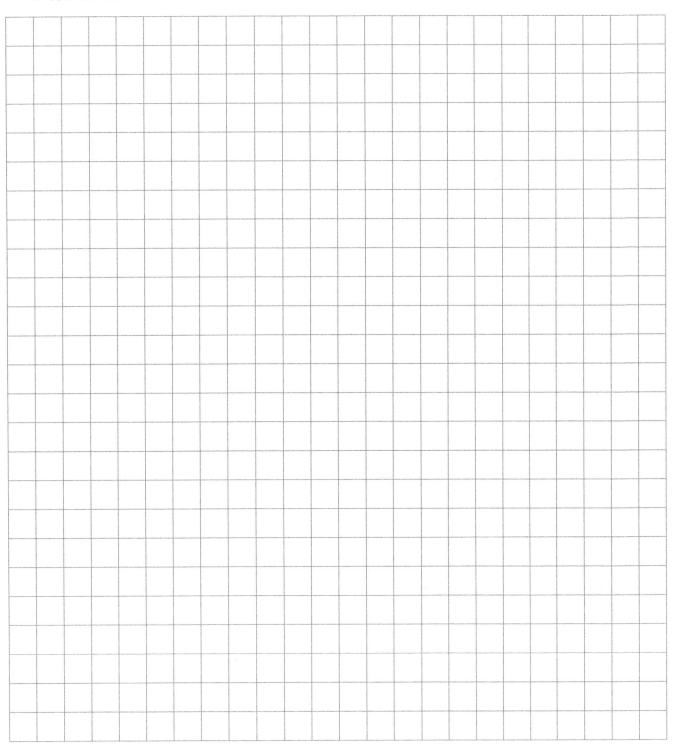

Problem 8. Markov needs to measure exactly 24 minutes for an experiment. He does not have an ordinary watch, but he does have two hourglasses. One of 20 minutes (Hourglass A), another of 6 minutes (Hourglass B). Will Markov be able to accurately measure the time required, how?

Olympiad 2013

(IV Mathematical Olympiad "Unikum")

Problem 1. How many five-digit numbers are there, whose sum of their digits is 2?

Problem 2. A computer multiplies a number by 2, then subtracts the number K from this result, then multiplies the result by 2 and subtracts K again, and so on. It performs each operation (multiplication and subtraction) 2013 times. Come up with a number that will not change as a result of the described work on the computer.

Problem 3. Cut the shape shown in the figure into six equal parts along the grid lines in all possible ways, if each of the resulting parts should contain one letter. Cutting methods are considered different if the parts obtained by one cutting method do not coincide when superimposed with the parts obtained by another method. Prove that there are no other ways besides those suggested by you.

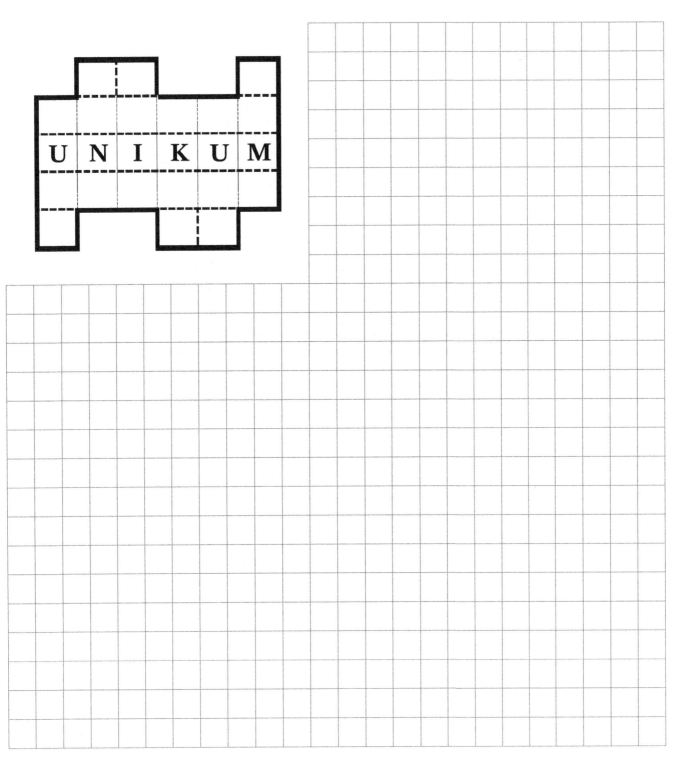

Problem 4. Petya had a basket full of apples. First, he met Masha and gave her half of his apples and half of another apple. Then he met Dasha and gave her half of his remaining apples and half of another apple. Then Petya lost half of his remaining apples and half of another apple. Finally, after he met Sasha and again gave her half of the remaining apples and half of another apple, the basket was empty. How many apples did Petya have at the beginning?

Problem 5. On the island of gentlemen and liars, gentlemen always tell the truth and liars only tell lies. Every inhabitant of the island is supposed to be a gentleman or a liar. Two of the three islanders A, B and C made the following statements:

A: "We are all liars".

B: "One of us is a gentleman".

Determine which of the three islanders A, B, and C is a gentleman and who is a liar.

Problem 6. Three friends decided to have a snack together, for this purpose one of them gave three sandwiches, the second - four sandwiches, and the third brought 70 rubles. How much of this money should the first take and how much should the second take so that the contributions of the three friends are equal? Let's consider all sandwiches equal and evenly divided.

Problem 7. Let us call a natural number "unique" if it does not change when the sheet on which the number is written is turned over (the lower and upper parts of the sheet are interchanged). Determine how many "unique" numbers there are among the four-digit numbers. In recording unique numbers, we will use only the numbers 0, 1, 6, 8, 9; examples of "unique" numbers: 1; 8; 69; 609.

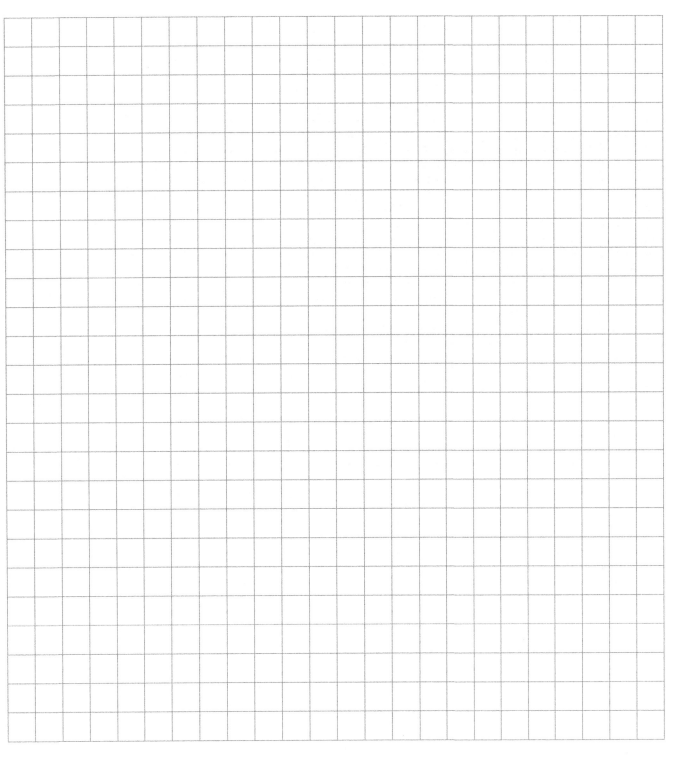

Problem 8. At his leisure, Kostya drew a labyrinth in which he encrypted the path from *Entrance* to *Exit* P as 2210. Unravel the code proposed by Kostya and determine which exit you can get if you use the path with code 3031?

Olympiad 2014

(V Mathematical Olympiad "Unikum")

Problem 1. For the number 248, the middle digit is two times different from the extreme ones. How many three-digit numbers are there in which the middle digit is two times different from the extreme ones?

Problem 2. Four grams of paint were used to paint a wooden cube. When it was dry, the cube was cut into 8 equal smaller cubes. How much paint is required to paint over the resulting unpainted surfaces?

Problem 3. Is it possible to bake a cake that can be divided into 5 parts by one straight cut? And into 2014 parts?

Problem 4. Schoolchildren Vasya, Petya, Irina and Masha took the first 4 places at the Mathematical Olympiad "Unicum". The sum of the places occupied by Vasya, Petya and Irina is 9, and the sum of the places of Irina and Masha is 5. What place did each of the named students take if Petya took a higher place than Vasya?

Problem 5. Mikhail drank 40 *ml* from a cup of coffee (a full cup contains 180 *ml*) and topped it up with milk. Then he drank another 60 *ml* from a cup of coffee with milk and again topped it up with milk. Then he drank half a cup and again refilled the cup with milk until it was full. Finally, Mikhail drank the full cup. What did he drink more: coffee or milk? How much?

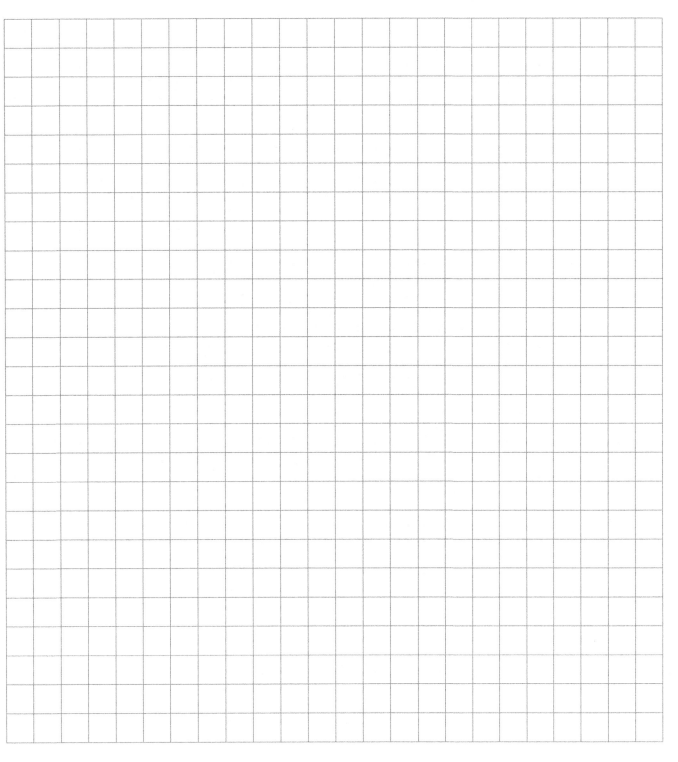

Problem 6. Murka the cat eats a tasty sausage in 6 minutes and Vaska the cat 2 times faster. How long will it take for them to eat the sausage together?

Problem 7. In the addition example, the numbers were replaced with asterisks. It turned out: ** + *** = ****. It is known that each of the terms and the sum will not change if you read them from right to left. Reconstruct the original example.

Problem 8. Vasya wrote natural numbers in an orderly fashion: 12345678910111213141516171819 ... Which digit is in the 2014th place?

Olympiad 2015

(VI Mathematical Olympiad "Unikum")

Problem 1. In the number 123456789101112131415 ... 20142015, crossed out all digits in even places; in the resulting number, all the digits in even places were crossed out again, and so on. The crossing out continued as long as there was something to cross out. What digit is left in the end?

Problem 2. Divide the quadrilateral shown in the figure into three parts with the help of two straight cuts so that a rectangle can be assembled from all the parts obtained. The parts must not overlap.

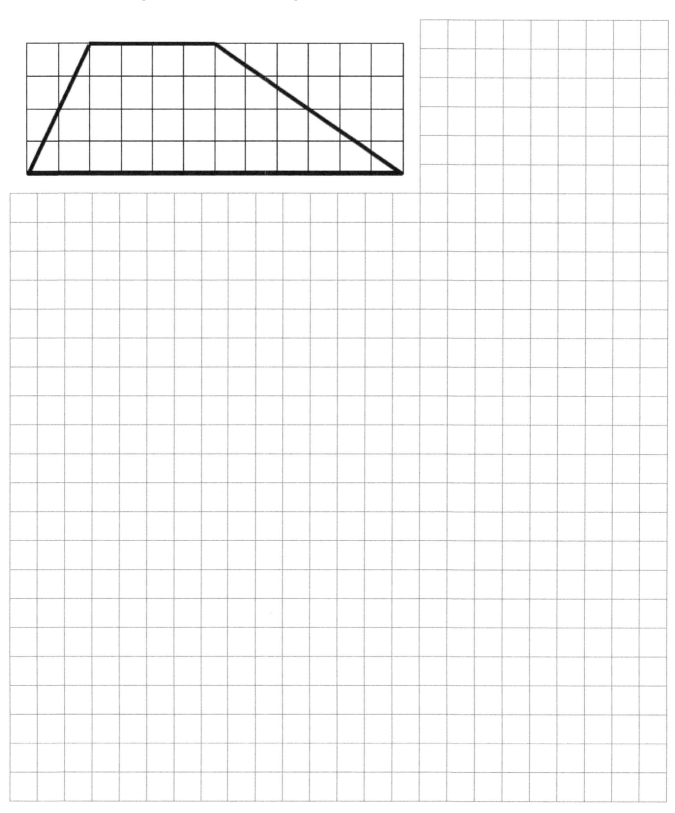

Problem 3. It is known that there are four Fridays and four Mondays in January. What day of the week is January 3rd?

Problem 4. For a given rectangle with sides of 3 *cm* and 2 *cm*, squares were constructed on each of its sides. Determine the area of the resulting shape.

Problem 5. Find all the solutions to the number puzzle MAY : AY = 25. Different letters represent different digits. Explain the lack of solutions other than those found.

Problem 6. (An old problem) A horse eats a cart of hay in a month, a goat in two months, a sheep in three months. How long will it take for a horse, goat, and sheep to eat the same cart of hay together?

Problem 7. Only 11 roses of red, blue and white flowers bloomed in the greenhouse (how many roses of each color are unknown). Every day there are more and more blooming roses: to each red rose one more red rose is added, to each blue one – three more blue ones, to each white one – five more white ones. Could the total number of roses be equal to 2015 in a few days, if during this period not a single rose has withered and has not been cut?

Problem 8. Two friends are playing a certain game. On the first move, the first player puts into a pile any number of stones less than 10. Further, the players alternately put into the pile any number of stones, but not exceeding the number of stones already in the pile. The one who brings the number of stones to 100 wins. Who wins – the first or the second? And how do you play to win?

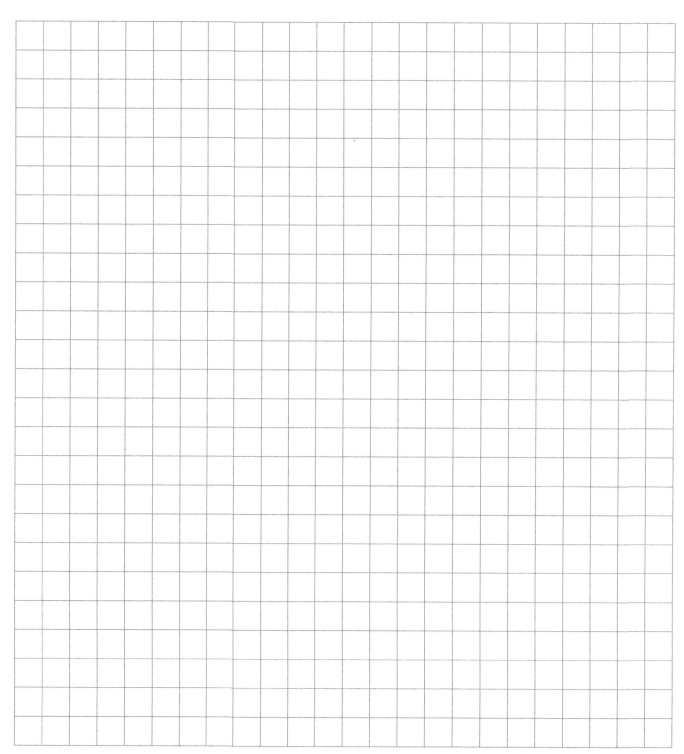

Olympiad 2016

(VII Mathematical Olympiad "Unikum")

Problem 1. During preparation for a math Olympiad, a student planned to solve 120 problems in 30 days, the same number of problems per day. However, every day he managed to solve two more problems than anticipated. How many problems did the student solve?

Problem 2. Can the numbers 2, 4, 6, 8, 2016 be written in a string so that one of the two adjacent numbers is divisible by the other?

Problem 3. What is the area of triangle ABC (see figure below) if the length of the side AC is 4 *cm* and the length of the side BC is 7 *cm*?

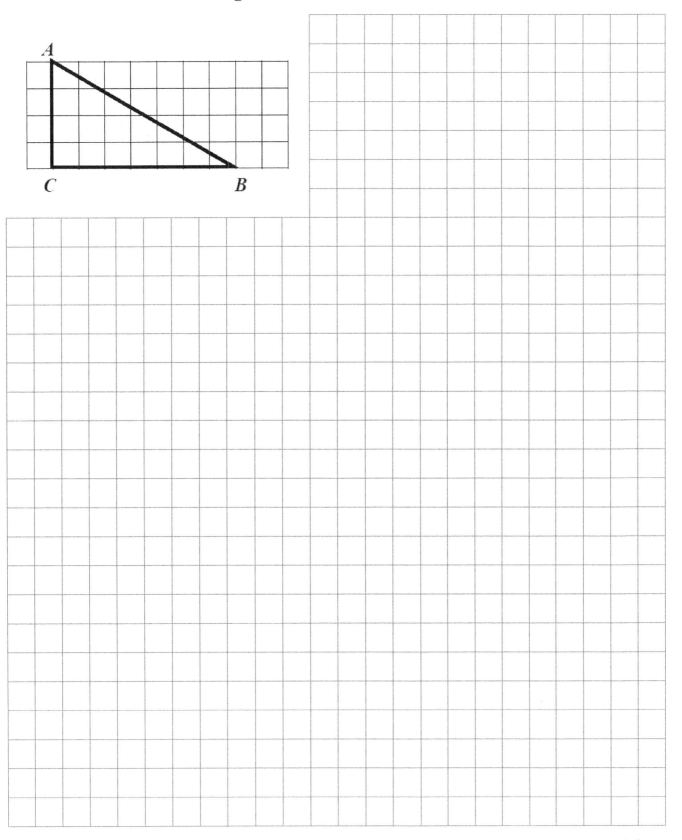

Problem 4. The Director of the Center "Strategy" handed out 2016 delicious sweets to the four students who showed the best results at the Olympiad. The fifth grader received two more than the sixth grader. A fourth grader received two more than the fifth grader, and a third grader received two more than the fourth grader. How many sweets did the third grader get?

Problem 5. Divide the strip into four equal parts (matching when overlapping) so that all the parts have the same sum of the numbers written in the cells.

2001	2009	2016	2007	2012	2005	2004	2003
2008	2015	2010	2002	2013	2006	2011	2014

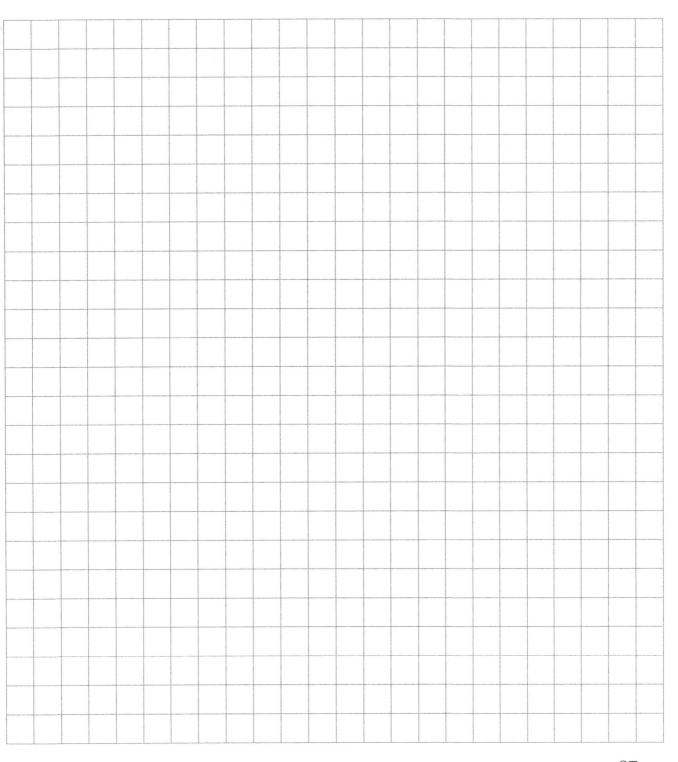

Problem 6. In the rectangle ABCD, the sides AB = CD = 5 *cm* and BC = AD = 10 *cm*. The points Q and T are the midpoints of the sides AB and CD. Points K and L are marked on side BC so that BK = 2 *cm* and BL = 5 *cm*. On the AD side, points M and F are marked so that AM = 2 *cm* and AF = 5 *cm*. Determine which quadrilateral has the largest area: QKTM or ALTF.

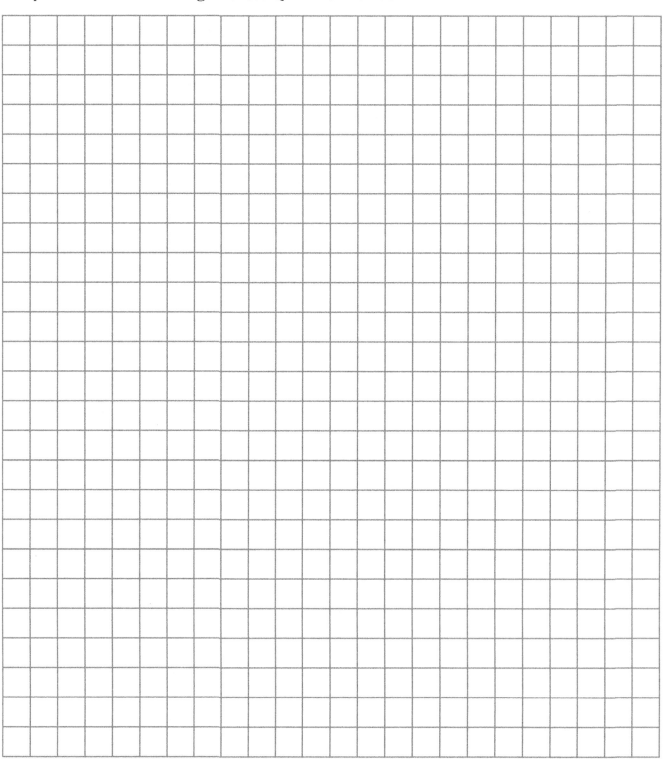

Problem 7. Polina and Vasilina came up with a game. They drew a picture on paper – three white flowers. Then the girls take turns repainting one flower, Polina begins. If the flower was white, it turns red, and if it was red, it turns white. On each move, the player can choose any flower (including the previously repainted one), but on condition that after changing the color, the picture does not become exactly the same as it was at some previous moment. The one who cannot make a move loses. Which player can guarantee herself a victory, no matter how her opponent plays?

Problem 8. Find all possible solutions to the puzzle: STRA + TEG – IU = 2016. Here, different letters represent different digits, and consonants correspond to digits no more than five, and vowels correspond to digits greater than five.

Olympiad 2017
(VIII Mathematical Olympiad "Unikum")

Problem 1. There are 17 coins on the table. At least one of them is worth 2 rubles. Whatever two coins you take, at least one of them will be worth 5 rubles. How much money is on the table?

Problem 2. Kid and Carlson poured 6 liters of jam into two jars and carried them home. Kid got tired and poured 2 liters of jam from his jar into Carlson's jar. From this, Kid's jam was half that of Carlson. How much jam did Carlson have initially?

Problem 3. A group of 23 schoolchildren decided to share the mushrooms they collected in the forest among themselves. If they divide the mushrooms from two baskets evenly (by quantity), there will be one mushroom left over. If the mushrooms from three baskets are divided evenly, there will be 13 mushrooms left over. What is the smallest amount of mushrooms in one of those baskets, if the amount of mushrooms in all baskets was the same?

Problem 4. Vasya and Petya love rings. However, Petya loves linked rings, but Vasya does not. Help Vasya determine the minimum number of rings to cut so that all five rings separate from each other?

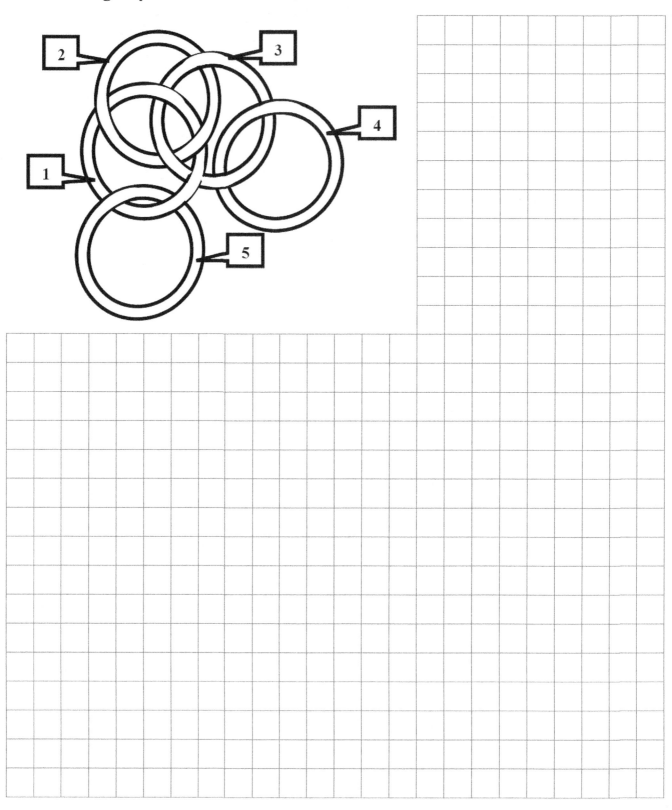

Problem 5. Gandalf has an unusual chessboard, it is smaller than usual: 5 by 5. How many ways can one or more cells be shaded on this board so that they form squares of different location or size? Gandalf managed to get 32 such squares, can you improve his result?

Problem 6. At a tavern in the town of Bree near Shire, visitors cast the roles of liars and knights for the night. The former always tell a lie, the latter always tell the truth. Thorin sat at a table with three local residents and asked each, "How many of your two friends are gentlemen?" To which he received the following responses. First: "None". Second: "One". What did the third say?

Problem 7. Hobbits Fili and Keely look at the aquarium and observe the trajectory of the fish, redrawing it on parchment. And they look at the adjacent walls of the aquarium. Fili depicted the trajectory on the right figure, Keely – on the left figure. What trajectory would Radagast have depicted, if he was watching from above?

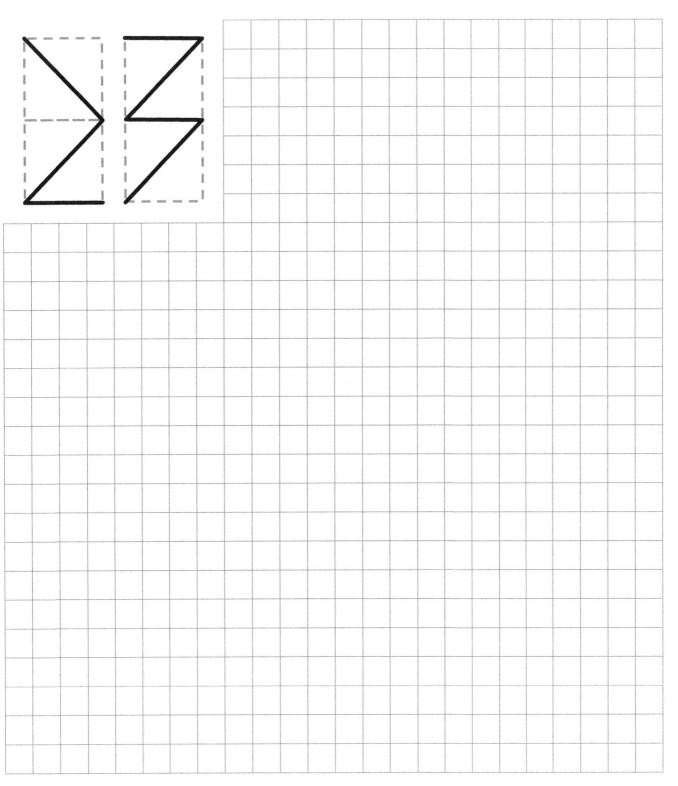

Problem 8. Ivan got acquainted with the division of the segment into parts. First, he marked points on the segment dividing it into 15 equal parts. Then he erased the marked points and drew points on the same line, dividing it into 25 equal parts. Then he again erased the marked points and drew points on the same segment dividing it into 35 equal parts. Then Ivan wondered how many parts the segment could be divided into if the points were not erased? Help Ivan solve this problem.

Olympiad 2018

(IX Mathematical Olympiad "Unikum")

Problem 1. Bilbo has a two-plate scale in his kitchen that allows him to compare the weights of the things put on the plates. The cherry pudding balanced 1 cupcake and 2 apples, while the pudding and apple balanced 2 cupcakes. How many apples will balance the pudding?

Problem 2. Losyash asked Kopatych to cut five logs 6 meters long each into half-meter pieces. How many times will Kopatych have to cut the logs, that is, how many cuts will he have to make?

Problem 3. Krosh, Nyusha and the hedgehog ate 5 carrots each, then they ate apples equal to twice as many carrots, then they ate pears equal to three times as many apples, and finally they ate candy equal to half the number of pears. How many different delicious items did Krosh, Nyusha, and the hedgehog eat together?

Problem 4. In the shape shown in the figure, you need to move three matches so that you get three squares.

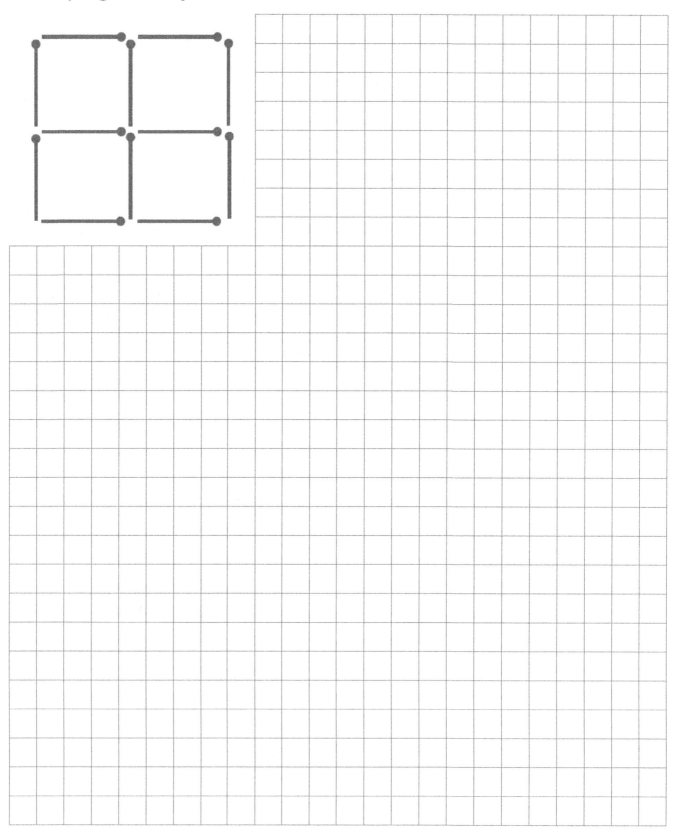

Problem 5. The Harvest Festival has come, and in the evening all the hobbits traditionally gather in the square to celebrate this day and watch the fireworks launched by Gandalf. It got dark outside, and Bilbo Baggins' house ran out of candles! And Bilbo, dressed in his best caftan, needs to get a pair of boots out of the trunk - blindly. He is known to have a total of 12 pairs of black and 12 pairs of blue boots. Of course, he would like to get a pair of black ones, but a pair of blue ones will do. He can take a certain amount of boots, bring it to the window, from where the street lamp is illuminating, and see what he took. What is the smallest number of boots he needs to take to get: a) a pair of the same color; b) a pair of black (all boots are mixed, and you also need to take into account that the pair should be left + right)?

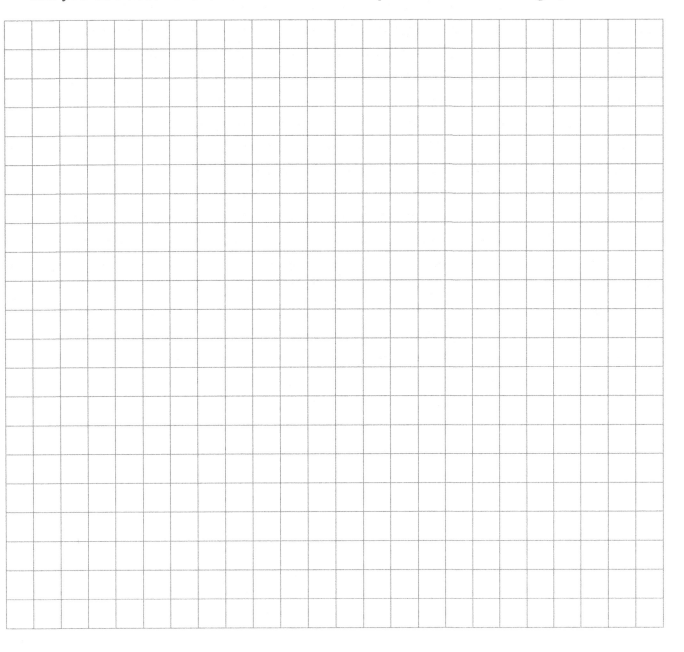

Problem 6. Losyash and Kopatych were fishing. Losyash caught 9 fewer perch than Kopatych, and Kopatych – 4 times what Losyash caught. How many perches did they catch together?

Problem 7. Twelve students of school № 777 came to the Mathematical Olympiad "Unikum". To refresh themselves before the competition, they bought seven identical pies. Help, if possible, divide the purchased pies equally among all 12 students, without dividing any of the pies into 12 equal parts.

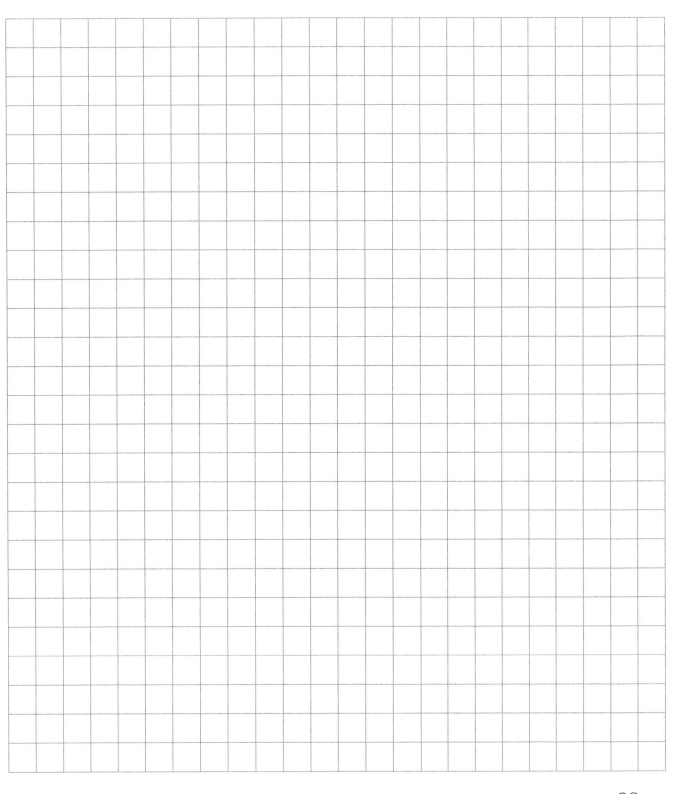

Problem 8. Bilbo Baggins was playing with his dominoes, but in the usual form this game did not please the hobbit, and he came up with another use for it.

a) Taking an 8×8 chessboard, Bilbo wants to fill it with 1×2 dominoes (without going over the edge of the board) so that only two dominoes form a 2×2 square. Help him do this.

b) Bilbo took an 8×8 chessboard and wants to pave (fill) it with 1×2 dominoes, leaving the lower left and upper right cells of the board free. Can Bilbo do it?

Olympiad 2019

(X Mathematical Olympiad "Unikum")

Problem 1. Two students encrypted numbers. The first drew a circle instead of an odd digit, and a square instead of an even one. The second student painted over the larger digit in each encrypted number. For example, the number 41 in encrypted form looks like this:  .

How many numbers they could encrypt in the form shown above?

Problem 2. Alexey climbed the stairs home and counted 176 steps. On which floor does Alexey live, if it is known that to climb one floor, he needs to go through 22 steps?

Problem 3. The clothing store sells 5 different shirts, 3 different ties and 4 different jackets. Sergei wants to buy two pieces of clothing with different names. How many options does he have?

Problem 4. Rick and Morty bought the same number of teleporters on the planet Squatch. How much does one teleport cost if Morty paid the cost of the teleport in 3 i.c.u bills (intergalactic currency units), and Rick – in 5 i.c.u bills, if they gave in total less than 10 bills to the cashier?

Problem 5. Rick formed the number 4352 by arranging 4 numbered cards he had (one digit per card: 2, 3, 4 and 5) and asked Morty to rearrange the cards to get a number greater than the original. He just so happened that in all the parallel worlds, all the Rick and Morty did the same, and all the numbers in all the worlds turned out to be different. What is the maximum number of parallel worlds that could exist (including the original world)?

Problem 6. Rick and Morty were captured by the belligerent but educated xenomorphs who respect the ability for logical, spatial and analytical reasoning. They gave the prisoners a 4 × 4 board, into which they put 4 stones. The condition of release - Rick and Morty need to divide the board into 4 equal parts (in shape and size), each of which will contain exactly one stone.

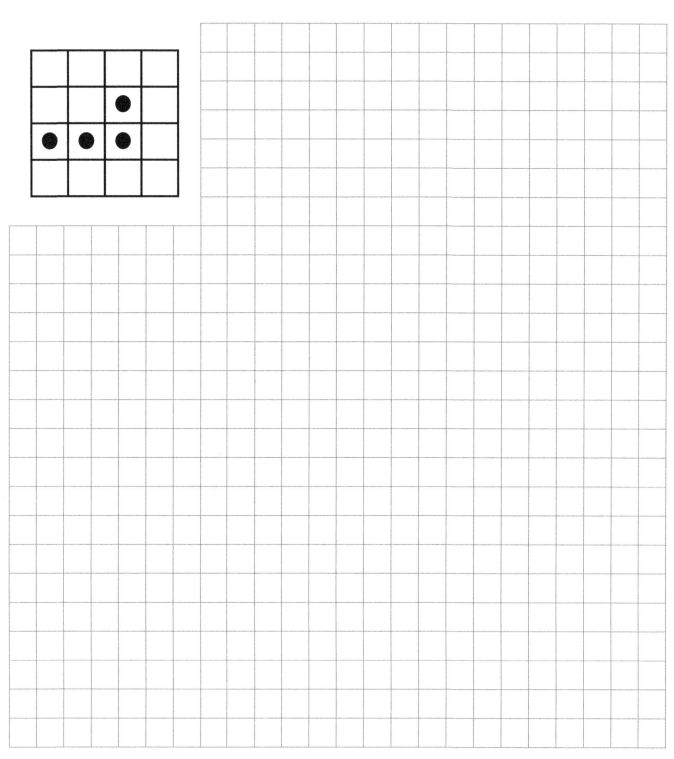

Problem 7. What is the smallest number of colors that can be used to color the cells of a 10 x 10 square, so that the cells of the same color have neither a common side, nor a common vertex.

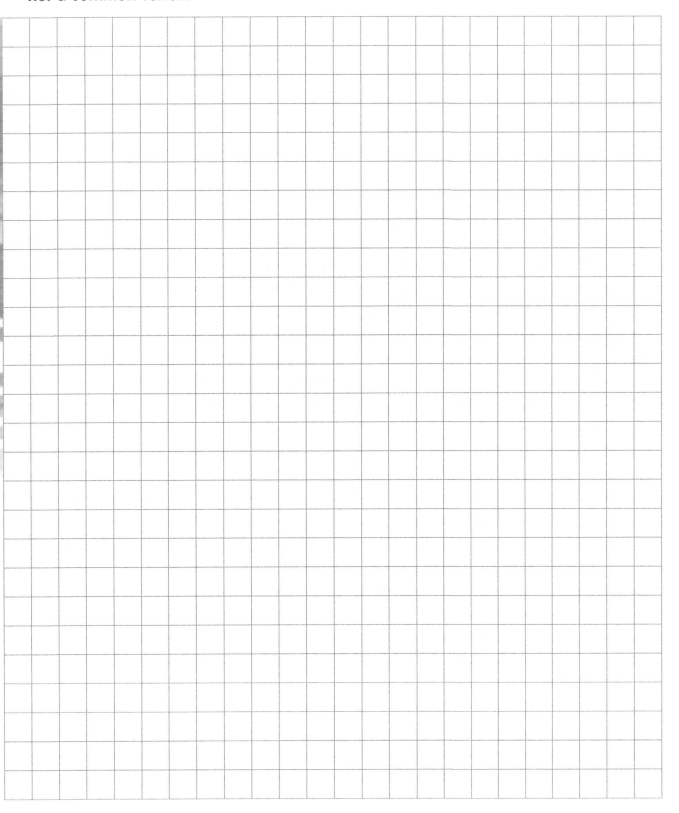

Problem 8. The snail travels on a cube suspended from a rope. The snail seeks to crawl along the shortest possible paths from some vertex A of the cube to the opposite vertex B. Find all the shortest possible paths of the snail.

Olympiad 2020

(XI Mathematical Olympiad "Unikum")

Problem 1. Six bushes were planted in the greenhouse: 1 orchid, 2 roses and 3 chrysanthemums. Every day new bushes were added to them: 3 orchid bushes, 5 rose bushes and one chrysanthemum bush. In how many days will there be 150 bushes of the indicated plants in the greenhouse in total? All plants took root well and did not dry out.

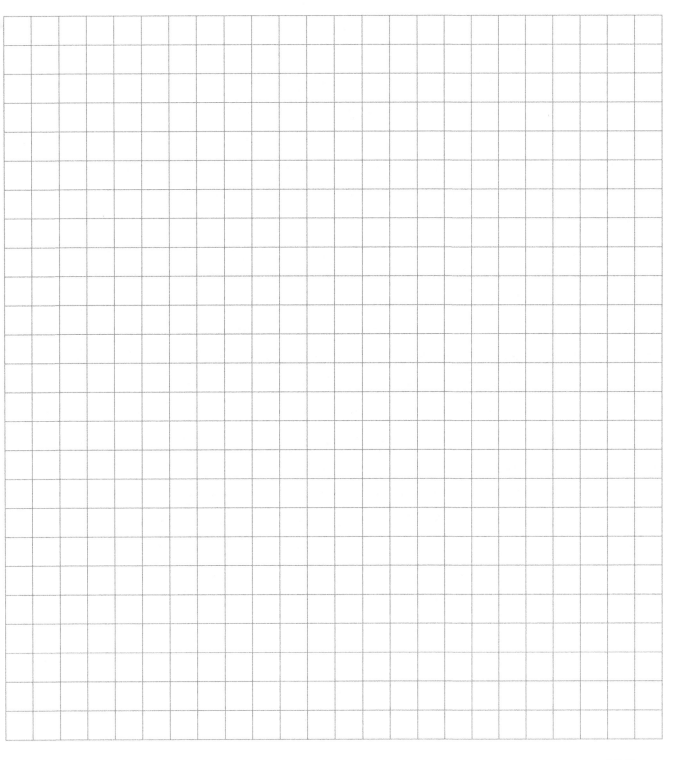

Problem 2. Use a few digits 1 (no other digits) and arithmetic operations to get the number 2020. You cannot use parentheses. Numbers can be made from digits.

Problem 3. Boris was presented with a toy consisting of 56 small wooden cubes (see picture). Boris drilled 6 through holes in the toy, each of which goes through exactly 6 small cubes. How many small cubes have three holes been drilled through?

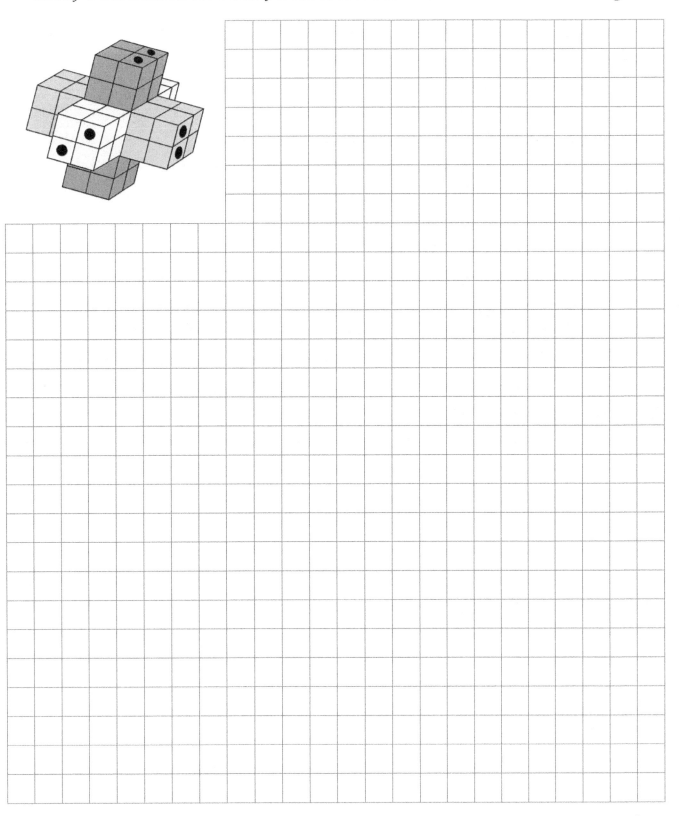

Problem 4. How many three-digit natural numbers are there, for which the middle digit is equal to the sum of the extreme ones?

Problem 5. Divide the shape shown in the figure into 12 parts so that the cut line goes along the sides of the cells, and there are 8 identical shapes of 4 cells, and 4 identical shapes of 5 cells. One of the necessary shapes is already in the figure.

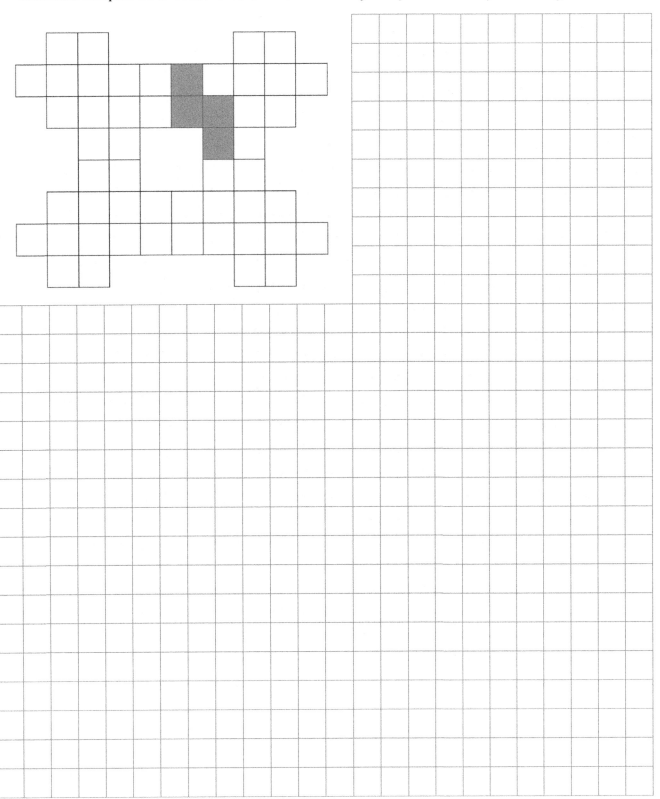

Problem 6. In the given 4 × 4 square, mark in each column, in each row and on each diagonal exactly one number so that the product of your chosen numbers equals 2020. On one diagonal are included the cells from the upper left corner to the lower right, and on the other – the cells from the lower left corner to the upper right.

4	5	101	1
5	2	101	1
2	4	5	5
10	1	4	2

Problem 7. In Unicumgrad, exactly three roads depart from each house. Each road leads from one house to another. Petya counted the number of roads in his hometown of Unicumgrad and counted 100 of them. Was he wrong? Why?

Problem 8. Smeshariki decided to celebrate the New Year. Krosh, Hedgehog and Barash were responsible for decorating the festive tree. Krosh brought two white balls for decoration, Hedgehog - two red ones, and Barash - two blue ones. Externally, the balls are the same size, but in each pair of the same color there is one light and one heavy ball. All light balls weigh the same among themselves, all heavy balls also weigh the same. When, half an hour after the start of work, Losyash came, he saw that Krosh, Hedgehog and Barash were arguing – which balls are light and which are heavy. Then Losyash proposed an experiment: he said that having a scale at hand, this problem is solved quickly. Help these friends to determine all light and all heavy balls with the help of two weighings on a two-plate scale.

Answers

Olympiad 2011

1. 550 rubles.

2. 23 bags.

3. 7 *kg*.

4. 24 different routes.

5. The final result is 7989.

6. Yes, an example is shown below,

	1		2		2	
1	1	1	2	2	2	2
	1	1	2	2	2	
1	1	1		3	3	3
	4	4	4	3	3	
4	4	4	4	3	3	3
	4		4		3	

7. The number 1.

8. U = 1, N = 2, I = 3, K = 4, M = 5, R = 9, E = 8, G = 7 and A = 0.

Olympiad 2012

1. The result ends with the digit 5.

2. First – 2, Second – 4, Third – 6.

3. 10 pencils.

4. 2 hours 58 minutes.

5. 1. If you got an orange, and there is no sign "Oranges" on the box, then we hang a sign "Oranges" on this box. On one of the remaining boxes we change the sign.
2. If you got an orange, and there is a sign "Oranges" on the box, then we hang a sign "Mix" on this box. On one of the remaining boxes we change the sign. The same works for apples.

6. Yes, it is possible. The figure below shows an example,

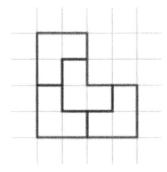

7. All solutions are shown below:
a) 10 + 24 + 10 + 25 + 30 = 99;
b) 20 + 14 + 20 + 15 + 30 = 99;
c) 10 + 25 + 10 + 24 + 30 = 99;
d) 20 + 15 + 20 + 14 + 30 = 99.

8. Yes, he will be able.

Answers

Olympiad 2013	Olympiad 2014

Olympiad 2013

1. 5 numbers.

2. The number K.

3. The figure below shows an example. More examples are possible.

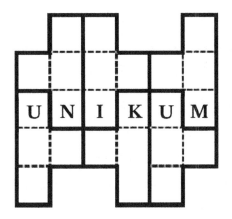

4. 15 apples.

5. B is a gentleman, A and C are liars.

6. 20 rubles – first, 50 rubles – second.

7. 20 numbers.

8. Second letter A.

Olympiad 2014

1. 12 numbers.

2. 4 grams.

3. Yes, it is possible.

4. Masha – First, Petya – Second, Vasya – Third and Irina – Fourth.

5. Milk, 190 *ml.*

6. 2 minutes.

7. 22 + 979 = 1001.

8. The digit is 7.

Olympiad 2015

1. The digit is 1.

2. An example is shown in the figure,

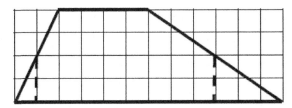

3. Thursday.

4. 32 cm^2.

5. 625 : 25 = 25.

6. 6/11 months.

7. No, it could not.

8. The first player.

Olympiad 2016

1. 180 problems.

2. Yes, for example: 6, 2016, 2, 4, 8.

3. 14 cm^2.

4. 507 sweets.

5. An example is shown below,

2001	2009	2016	2007	2012	2005	2004	2003
2008	2015	2010	2002	2013	2006	2011	2014

6. The areas are equal.

7. Polina.

8. All solutions are shown below,
a) 1529 + 573 − 86 = 2016 (S = 1; T = 5; R = 2; A = 9; E = 7; G = 3; I = 8; U = 6),
b) 1538 + 574 − 96 = 2016 (S = 1; T = 5; R = 3; A = 8; E = 7; G = 4; I = 9; U = 6),
c) 1539 + 564 − 87 = 2016 (S = 1; T = 5; R = 3; A = 9; E = 6; G = 4; I = 8; U = 7).

Olympiad 2017

1. 82 rubles.

2. 2 liters.

3. 12 mushrooms.

4. 2 rings.

5. 55 ways.

6. One.

7. The figure below shows the trajectory,

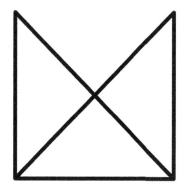

8. 65 parts.

Olympiad 2018

1. 5 apples.

2. 55 times.

3. 180 items.

4. An example is shown in the figure:

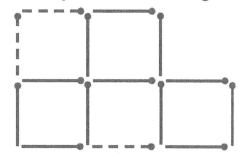

5. a) 25 boots; b) 37 boots..

6. 15 perches.

7. 1) Each student's share is 7/12 of the pie. 2) 7/12 = 1/3 + 1/4. 3) First divide each of the 4 pies into three equal parts, we get 12 parts of 1/3 of the pie. Then, divide each of the 3 remaining pies into four equal portions to make 12 slices of 1/4 pie. Having given each student 1/3 and 1/4 of the pie, we get the solution to the problem. *Note.* Dividing the pie into parts divisible by 12 does not satisfy the condition of the problem.

8. a) The figure below shows an example. b) It is not possible.

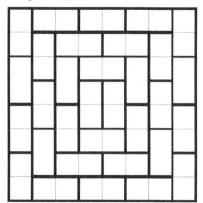

Olympiad 2019

1. 10 numbers.

2. On the 9th floor.

3. 47 options.

4. 15 i.c.u.

5. 9 (taking into account the original world).

6. An example is shown in the figure:

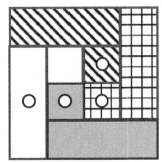

7. 4 colors.

8. 6 paths.

Olympiad 2020

1. After 16 days or 17 days (if, as the first day, the day when 6 bushes were planted was counted).

2. For example, 1111 + 1111 − 111 − 111 + 11 + 11 − 1 − 1.

3. 1 small cube.

4. 45 numbers.

5. See the next figure:

6. An example is shown below:

4	5	101	1
5	2	101	1
2	4	5	5
10	1	4	2

7. He was wrong.

Answers

Made in the USA
Las Vegas, NV
24 May 2023